Energy 134

更多氧气，更多生命力

More Oxygen, More Life

Gunter Pauli

[比]冈特·鲍利　著

[哥伦]凯瑟琳娜·巴赫　绘

贾龙智子　译

上海远东出版社

目录

Contents

一条沙蚕正忙着在沙子里蠕动。一只小鹬在一旁仔细地看着沙蚕身后留下的弯弯曲曲的痕迹。

"你在忙着掘沙……是在埋一些不想让我看见的东西吗？"这只姬滨鹬问道，在所有鹬鸟中，它是体型最小的。

$\mathcal{A}$ sea worm is busy burrowing in the sand. A small sandpiper is closely observing the curly structures the sea worm leaves behind.

"$\mathcal{Y}$ou are so busy digging ... are you burying something you do not want me to see?" asks the smallest of all the sandpipers, the Least Sandpiper.

一条沙蚕正忙着在沙子里蠕动。

A sea worm is busy burrowing in the sand.

谁会吃沙子呢？

Who eats sand?

"我不是在藏东西，我是在吃东西。"沙蚕回答道。

"谁会吃沙子呢？"

"嘿，你根本不知道混在沙子里面的食物有多么丰富。没谁知道这事，所以我就独吞了。"

"I am not hiding anything. I am eating," responds Sea Worm.

"Who eats sand?"

"Well, you have no idea how rich the food is that is mixed in along with the sand. Nobody knows about it, so I have it all to myself."

"你看起来一辈子都是头冲下颠倒着过！你不会头痛吗？"小鹬鸟问道。

　　"从来不会。而且我有很好的理由倒立过来——我可不想在上厕所的时候毁了自己的食物！"

"You seem to live your entire life with your head down, and your bottom up! Don't you get a headache?" Sandpiper asks.
"Never. And I have a good reason to keep my bottom up – when I go to the loo, I don't want to soil my food!"

......头冲下颠倒着过！

... head down, and your bottom up!

……极好的水利工程师吗？

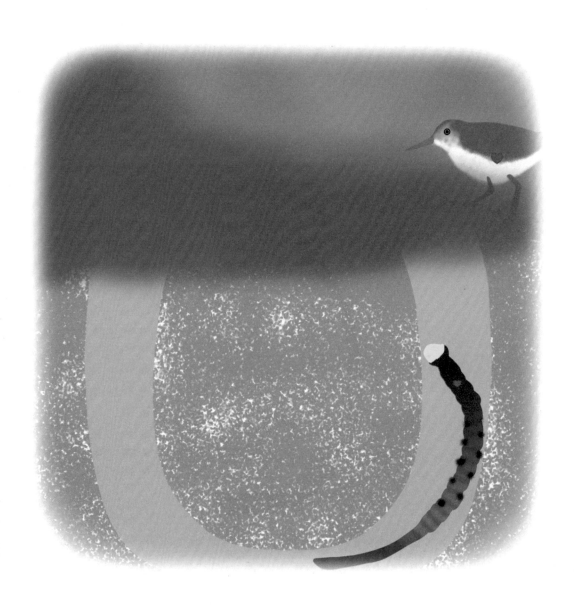

... great water engineers?

"不好意思，但我有另一个问题：海水一直冲刷着你的头，你怎么呼吸呢？"

"你不需要担心我。你不知道我们沙蚕是极好的水利工程师吗？而且不止这些——我们是最早开始利用血红蛋白的生物之一。"沙蚕自夸道。

"Excuse me, but I have another question: How do you breathe when the sea water washes over your head all the time?"

"You need not worry about me. Don't you know that we sea worms are great water engineers? And that's not all – we were some of the first creatures to ever make use of haemoglobin," Sea Worm boasts.

"血……什么？"

"血红蛋白！别跟我说你从来没听过这个词——怎么会有这么无知的家伙？"沙蚕呵斥小滨鹬道。

"很抱歉，我确实没听说过。"胆小的小鹬鸟坦白道。

"Haemo-what …?"

"Haemoglobin! Don't tell me that you have never heard of it – how ignorant can one be?" Sea Worm barks at Sandpiper.

"I am sorry to say I haven't," confesses the timid little Sandpiper.

血……什么?

Haemo-what ...?

......冷血动物还是温血动物。

... cold or warm-blooded.

"你知道，你也有的。所有的动物都有这个东西，不管它们是冷血动物还是温血动物。"

"叫什么来着？我们都有的，名称很复杂的——血……什么？"

"血红蛋白。我的错，但是你也太笨了！你难道连只有四个音节的词都记不住吗？血—红—蛋—白！"

"You have it as well, you know. All animals have it, whether they are cold or warm-blooded."

"What is it again? That thing we all have, with that difficult name – haemo ...?"

"Haemoglobin. My mistake, but you are stupid! Can't you even remember a word with only four syllables? Hae-mo-glo-bin!"

"我很抱歉，我可能年纪太小，这个词对我来说太复杂了。但你介意告诉我它的用处吗？"

"现在听好了，我要解释了。"沙蚕说道。

"你看，我吃东西的时候头埋在我挖的坑里，我需要氧气提供能量。由于周围有沙子和水，有时我很难得到空气——所以我把氧气储存在我血液里的血红蛋白中。"

"I'm sorry, I may be a little too young for such big words. But would you mind telling me what it is good for?"

"Now listen carefully and I will explain," Sea Worm says.

"You see, when I am eating, with my head down in my burrow, I need oxygen for energy. With all that sand and water around, it is sometimes hard to get air – so I simply store oxygen in my haemoglobin, in my blood."

......把氧气储存在我血液里的血红蛋白中。

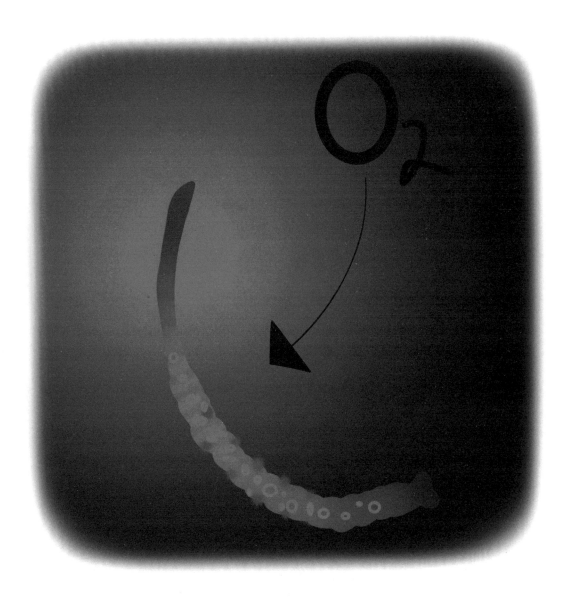

... store oxygen in my haemoglobin, in my blood.

......可以起死回生。

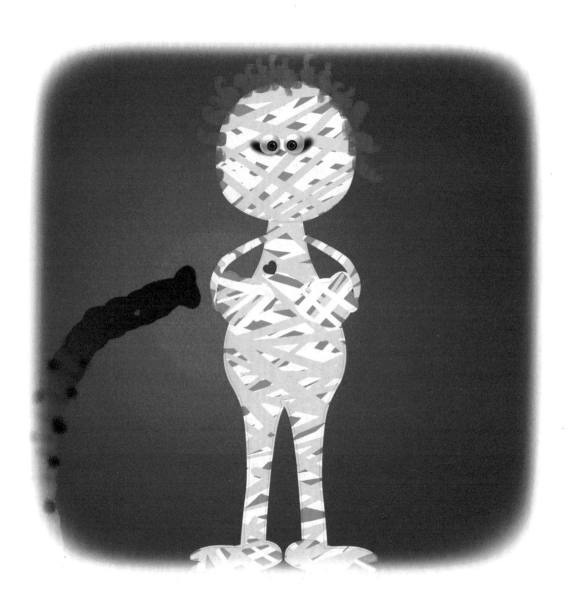

...can bring the dead to life again.

"那你为什么不把它就叫作血液呢？"

"因为它只是血液中的一部分，傻瓜！听着，提到储存氧气，我可是世界上最厉害的选手之一。我有这么多的血红蛋白，我的血可以起死回生。"

"听上去不大可能……"

"So why don't you just call it blood?"

"Because it forms only one part of blood, silly! Look here, when it comes to storing oxygen I am one of the world champions. I have so much of it that my blood can bring the dead to life again."

"That sounds impossible ..."

"别再说不可能了！看着我——我不但能通过皮肤呼吸，屏住呼吸数小时，而且只要我得到足够的食物，我的血液就能救别人的命。事实就是如此。"

"沙蚕先生，我真的很佩服你这一点，希望有一天我能像你一样对世界有所贡献……而且跟你一样友好和谦逊。"

……这仅仅是开始！……

"Don't you ever say that something is impossible! Look at me – not only can I breathe through my skin and hold my breath for hours, but my blood can save the lives of others – as long as I get enough to eat, that is the truth."

"I really admire you for it, Mr Sea Worm Sir, and wish that I can one day also be as useful to the world as you are … and as friendly and humble."

… AND IT HAS ONLY JUST BEGUN! …

...... 这仅仅是开始！ ......

... AND IT HAS ONLY JUST BEGUN! ...

The sea worm lives in a U-shaped shaft lined with mucus. It moves forward with a piston-like movement, while keeping the sand loose through careful blending with water entering through waves of contraction of the body.

沙蚕居住在内壁附有黏液的U形洞穴里。它们像活塞一样向前移动：通过身体一波波地收缩带动水流，仔细地将沙子与水流混合而使沙子松动。

There are up to 150 sea worms in one square meter of coastline. Commercial worm catchers in Brittany (France) remove up to 20 million worms a year for fishing, which is estimated to be less than 0.75% of the total population.

在一平方米的海岸线里有多达150条沙蚕。法国西北部布列塔尼地区的商业沙蚕捕手一年里为了捕鱼要清除多达2 000万条沙蚕，这一数字估计还不到沙蚕总数的0.75%。

The sea worm stores its faeces in its tail, reducing the need to expel it. Should a fish bite off the tail, then the remaining segments of the sea worm will grow bigger.

沙蚕把其排泄物存储在尾部以减少排出的需要。要是一条鱼咬掉了沙蚕的尾巴，那么沙蚕余下的部分会长得更大。

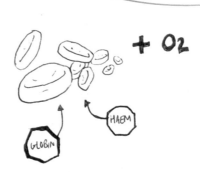

Haemoglobin combines with oxygen, enabling blood to carry 70 times more than if oxygen were simply dissolved in blood. The haemoglobin consists of part iron (haems) and part protein (globin).

血红蛋白结合氧气，使血液中所携带的氧气量比起仅仅溶解在血液中时的氧气量高70倍。血红蛋白是由铁（血红素）和蛋白质（球蛋白）构成的。

Healthy people have 150 grams of haemoglobin per litre of blood, which can bind with 200 ml of oxygen per litre. One quarter is distributed through blood, while three quarters return to the lungs.

健康的人体每升血中有150克血红蛋白，可结合200毫升的氧。四分之三的氧回到肺部，而四分之一的氧通过血液输送到身体各处。

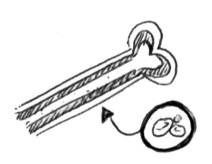

Haemoglobin is contained in red blood cells, and gives blood its red colour. The red blood cells are made in bone marrow.

血红蛋白包含在红细胞里，赋予了血液鲜红的颜色。红细胞是在骨髓中制造出来的。

$$\bigcirc = 150 \times O_2$$

There is 150 times more oxygen in a sea worm's haemoglobin than in that of a human. This opens the way for better preservation of organs for transplants.

沙蚕的血红蛋白中捕捉的氧气比人类血红蛋白中的氧气高150倍。这为待移植器官提供了更好的保存方法。

The Least Sandpiper is one of migratory birds that visit Brittany, a region rich in biodiversity, where sea worms thrive along the sandy coast.

姬滨鹬是逗留在法国布列塔尼的迁徙鸟类之一。布列塔尼地区生物多样性丰富，那里的沙质海岸中沙蚕数量庞大。

Would you be a better sportsman if you have more oxygen in your blood?

如果你的血液中有更多氧气，你会成为更优秀的运动员吗？

If haemoglobin makes up only 0.6% of the weight of the worm, what would you do with the rest of the worm?

如果血红蛋白只占沙蚕重量的0.6%，你会用沙蚕剩下的部分做什么？

Would you like to keep a sea worm as a pet?

你想拿沙蚕当宠物养吗？

Does eating sand sound appetising to you?

食用沙子听上去会让你产生食欲吗？

If you ever have the opportunity to go to a specialty fishing store, you may find sea worms for sale, if it is the right season for them. Ask if you may study a few of the worms closely. Hold them in you hands and carefully observe the way they are built. Find out if the people working in the store know about the sea worm's unique ability to store oxygen. Tell them all about it and share facts about the sea worm's incredible contribution to saving organs for transplanting. Do you think they could ever have imagined that something used by fisherman for bait could play such a unique role in the lives of people?

如果你有机会去一家专业的渔具商店，你可能会在合适的季节找到供出售的沙蚕。向店家询问是否可以让你近距离研究几条沙蚕。把它们放在手里，仔细观察它们的构造。询问店里的工作人员，看他们是否了解沙蚕独特的储存氧气的能力。告诉他们这些知识，并分享沙蚕对于保存待移植器官的惊人贡献。你觉得他们有可能设想过渔夫用来当鱼饵的东西能够在人类生活中发挥如此独特的作用吗？

学科知识
Academic Knowledge

生物学	沙蚕（也叫海蚯蚓）种类达1 000种，属于环节动物门；沙蚕是捕食者，某些沙蚕是寄生生物；沙蚕在空气和水里都能呼吸；沙蚕没有血型；姬滨鹬是最小的滨鸟和迁徙者。
化 学	血红蛋白和叶绿素的区别是分子核心的铁和镁原子；含盐的水和海水；氧气通过身体向器官运送；用盐水输液来补充血容量；柠檬酸钠可用作血液的抗凝剂，现已被枸橼酸葡萄糖取代。
物 理	沙蚕肌肉的收缩和展开让其能够吃下两倍于其体型的动物；沙蚕能将肌肉收缩10倍；沙蚕能生成一种滑腻的黏液，使其在沙子上挪动的过程中减少摩擦；储存沙蚕的血液数年；用于移植的器官是冷藏而非冷冻；沙蚕有156个氧结合位点，人类仅有4个。
工程学	活塞；如何在不损失氧气的情况下将携带氧气的血液转化成粉末；血液最初是保存在真空瓶子里的；搏动灌注设备保持液体持续涌入以移除人体用于移植的肾脏；沙蚕血红蛋白中储存的氧能够逆转脑缺氧。
经济学	延续器官的生命力及发现与病人完美相配的器官对社会的价值、成本的降低和生命质量的提高；血液的储存，输血的需求；产量提升改变了市场；如果沙蚕的任意部位被切断之后都可以重新长成一只沙蚕——在一只15厘米长的沙蚕的整个生命周期里它能产生超过20万条沙蚕。
伦理学	通过移植死去的人的器官有机会挽救生命；人类器官的非法交易。
历 史	威廉·哈维在1628年建立血液循环理论；卡尔·兰德斯坦纳在1900年发现了3种血型；1930年列宁格勒（今圣彼得堡）设立了第一家用于输血的血库；第一例人体器官移植的是肾脏，完成于1954年，第一例心脏移植完成于1968年。
地 理	沙蚕经常出现在潮水冲刷的海滩；沙蚕的生存范围分布太平洋和大西洋沿岸。
数 学	计算来自优化系统而非最大化某一参数所带来的多种收益：使用系统动力学来论证更完美匹配的移植器官，利用沙蚕的废弃物来降低海鲑鱼养殖的成本。
生活方式	为了增加血液中的氧气水平，人类需要锻炼、呼吸新鲜空气、喝更多的水、用适当的方式呼吸、减少盐分摄入和避免饮酒；富含铁的食物可预防缺铁性贫血，增加血液携氧能力。
社会学	对拥有更长寿命的渴望和社会对移植器官的接受程度上升，共同提高了生命的质量和长度。
心理学	接受器官移植后的挫败、困惑和负罪感。
系统论	从沙蚕体内获取移植器官的保存剂的高成本，被利用沙蚕其他部位当作鲑鱼的饲料等途径而抵消了。

教师与家长指南

情感智慧
Emotional Intelligence

小鹬鸟

小鹬鸟对沙蚕的行为有所质疑，有把握地问其是否在隐藏某些东西。沙蚕的解释并没让小鹬鸟满意。尽管鹬鸟体型很小，它有勇气维护自己。小鹬鸟想了解沙蚕是如何头朝下在水和沙中生存的。它不知道"血红蛋白"一词并在沙蚕辱骂它缺乏知识的时候承认了这一点。即使沙蚕继续用很苛刻的方式对待它，小鹬鸟为了搞明白更多关于这个主题的知识，始终保持友好和谦逊的语气。当沙蚕解释其独特的血液性能时，小鹬鸟一开始表示了怀疑，之后间接表达了对沙蚕不友善行为的惊讶（称其友好和谦逊其实是说反话）。

沙　蚕

沙蚕的态度很直接。它很自大，而且它给的答案很简略。沙蚕对自身有很高评价，认为自己是合格的水利工程师。当小鹬鸟不知道血红蛋白是什么的时候，沙蚕的回应是羞辱它并说它蠢。当鹬鸟不受沙蚕苛刻语气的影响，仍然温和地坚持要答案的时候，沙蚕最终解释了它血液的作用以及血红蛋白和血液的区别。无论如何，因为沙蚕非常了解自己的特性，整个对话过程中它都保持自信到自负的程度。沙蚕对鹬鸟关于态度的微妙评论并不敏感（鹬鸟说它想像沙蚕一样友好和谦逊，即便沙蚕完全不是这样）。

艺术
The Arts

找一些沙蚕的照片或到它们生活的沙滩去，仔细看看它们如何用独特的方式将自己的废弃物从藏身处堆积到成堆的沙子里。现在让我们也用泥来创造一些结构和形态。我们把这叫作泥艺。比一比，看谁最有创造力？谁能创造出最高的坚固的泥塑？

思维拓展
Systems: Making the Connections

现代科技让我们得以取得很多进步。器官移植给之前承受病痛或已面临死亡的人带来了健康上的根本性改变。通过移除病变的器官并替换来自他人的健康器官，病人得到了延长寿命的机会。器官移植成功的关键是良好的配对，搜寻配对的过程可能会很复杂。离开身体后的器官只能保存数小时，这一短暂的时间限制了理想配对。不够完美的配对需要更多的免疫抑制药物等手段。

沙蚕储存氧气的能力因而成为及时的革命性发现，其所携带的氧是人类血液的150倍。从一吨沙蚕里只能提取出6升的保存液。一些经济学家因此认为这种液体过于昂贵，可能每吨沙蚕只能为3个移植的器官提供保存液。然而，一套系统方法给出了完全不同的观点。沙蚕是渔夫首选的鱼饵之一，渔夫很快指出在提取保存液之后，海鳟鱼会很乐意吃掉剩余的沙蚕。这意味着器官移植液的生产同时为养殖渔业提供了丰富的饲料。沙蚕的养殖因此产生了两个产品和两个机会：高质量的海鳟鱼养殖，以及对移植器官进行更好地保存。

现在产生的问题是：这两种商业模式中哪一种支付了间接成本？人们认为海鳟鱼饲料的售价应该很高，而保存液的成本较低。事实上，海鳟鱼养殖产生的收益非常充分，甚至可能用于提供免费的器官移植所需的保存液。布列塔尼大西洋沿岸港口的老渔业随着捕捞下降导致渔业活动减少，那里的大量空间将能够得以利用。这些海港现存的基础设施可以用来养殖沙蚕和海鳟鱼。这将降低养殖的建设成本，让整个操作过程更具竞争性。由于沙蚕主要的食物来源是长在砂粒上的细菌和酵母，当地的有机啤酒酿造公司有多余的酵母可充当沙蚕的额外饲料。这种系统的商业模式重塑了食物链，同时重塑了医疗保健制度的某些方面，所以器官移植的保存液某天可能会免费提供！

动手能力
Capacity to Implement

血液是独特的身体产物。沙蚕血红蛋白的发现提供了一种血液替代品并给我们带来了惊喜。然而为了确保获得充足的血红蛋白量，需要更多的生产制造中心。应该在哪里生产？如何进行？在尝试通过更成功的器官移植来挽救人类生命的同时，过度捕捞和可能使沙蚕灭绝的风险是什么？现在起草一份行动方案，满足减少生态风险和金融风险的需求。跟朋友们讨论你的方案。

故事灵感来自
This Fable Is Inspired by

弗兰克·扎尔
Franck Zal

　　弗兰克·扎尔，1966 年生于法国巴黎，最开始研究的是海洋生物学。1996 年，他以血红蛋白为主题的论文获得巴黎第四大学的博士学位。弗兰克之后分别在加利福尼亚大学圣芭芭拉分校和比利时的安特卫普大学进行了一年的博士后研究。1992 年在法国的布列塔尼，他发现了沙蚕的独特能力，但直到 2006 年他才放弃法国国家研究中心的稳定工作，转而成立自己的公司。他持续开发独特的商业服务，为输血提供唯一的替代品并挽救脑损伤病人。他成功地筹集到足够的资金，在沙蚕繁荣了几百万年的地区制造出用于医疗卫生的创新性产品。

图书在版编目(CIP)数据

冈特生态童书.第四辑:修订版:全36册:汉英对照 /
(比)冈特·鲍利著;(哥伦)凯瑟琳娜·巴赫绘;
何家振等译. —上海:上海远东出版社,2023
书名原文:Gunter's Fables
ISBN 978-7-5476-1931-5

Ⅰ.①冈… Ⅱ.①冈… ②凯… ③何… Ⅲ.①生态环
境–环境保护–儿童读物—汉、英 Ⅳ.①X171.1-49

中国国家版本馆CIP数据核字(2023)第120983号
著作权合同登记号图字09-2023-0612号

策　划 张　蓉
责任编辑 曹　茜
封面设计 魏　来 李　廉

冈特生态童书
更多氧气,更多生命力
[比]冈特·鲍利　著
[哥伦]凯瑟琳娜·巴赫　绘

贾龙智子　译

记得要和身边的小朋友分享环保知识哦!
八喜冰淇淋祝你成为环保小使者!